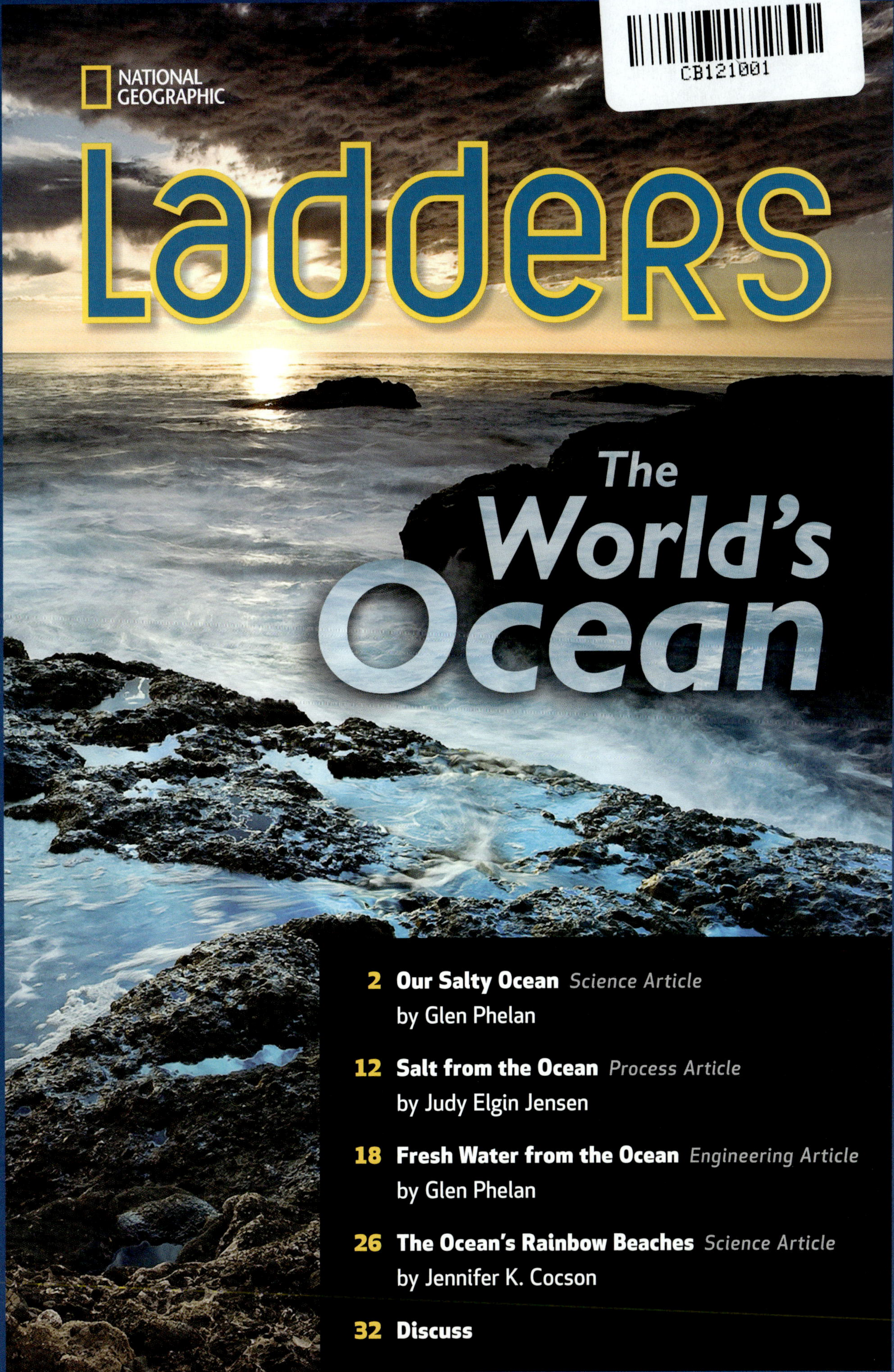

NATIONAL GEOGRAPHIC

Ladders

The World's Ocean

GENRE Science Article

Read to find out about properties of the vast world ocean.

Our Salt

Imagine standing on a beach and looking out to sea. The ocean stretches to the distant horizon with more water than you have ever seen. Yet this is only a tiny fraction of Earth's vast ocean. The continents separate the ocean into the Arctic, Atlantic, Pacific, and Indian Oceans. However, they are all connected into one global ocean. The ocean is an amazing place full of life, mystery, and surprises.

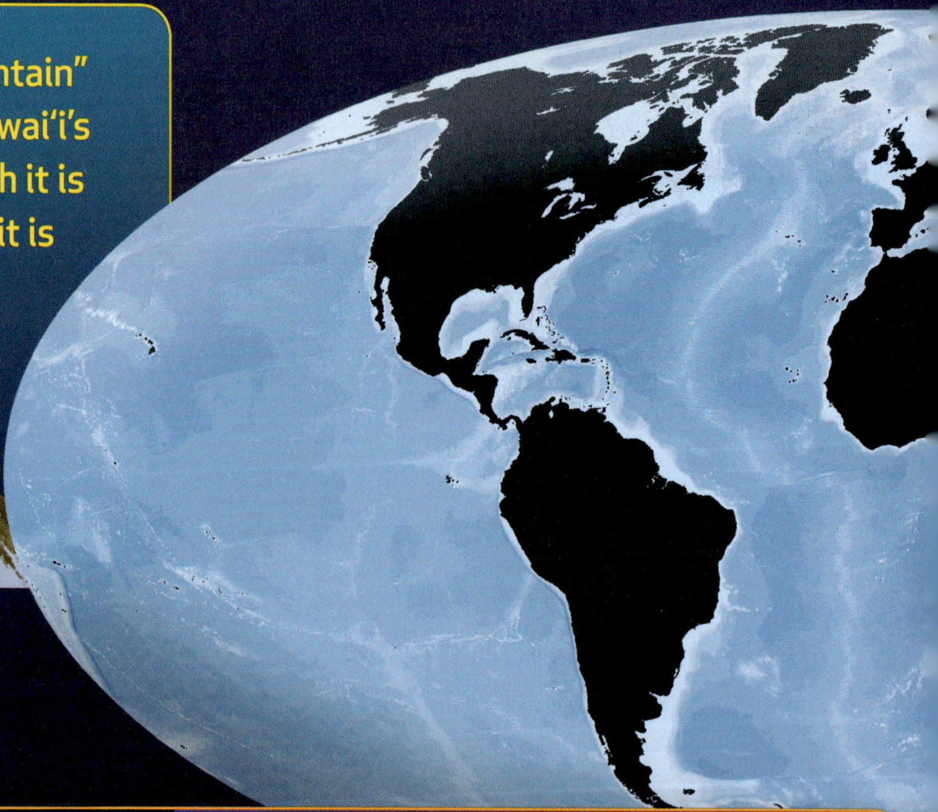

Earth's "tallest mountain" honor goes to . . . Hawai'i's Mauna Kea. Although it is mostly underwater, it is actually taller than Mount Everest.

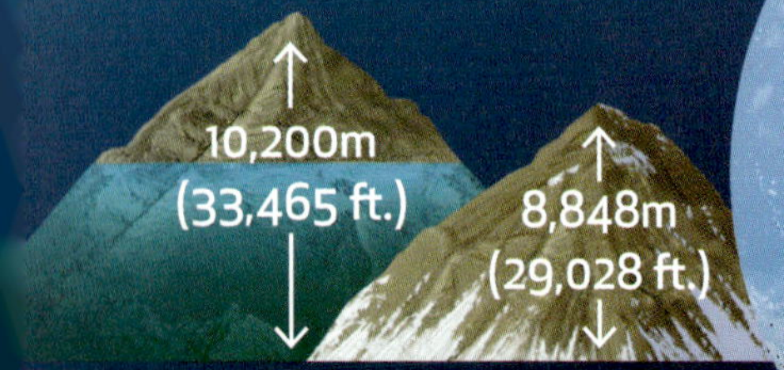

Most of the ocean is so deep that the sun's light cannot reach its depths. Angler fish that live in this darkness have glowing lures that attract prey.

y Ocean

by Glen Phelan

Ocean water supports the huge mass of the blue whale. Earth's largest animal ever, it stretches to a length of about 2.5 school buses.

A milk jug's volume of ocean water contains 9 large spoonfuls of salt.

Salt Water

Have you ever gone swimming in the ocean? All it takes is a little water on your lips to really taste the salty seawater! Or maybe you've felt a sting as the salty water got into a cut on your skin.

The sting comes from the salt in the water. Ocean water is mostly a **mixture** of water and different kinds of salts. The salts **dissolve** or break up into particles that spread evenly throughout the water, which makes salt water a kind of mixture called a **solution**. If the ocean's salts could be removed and spread evenly over the continents, they would form a layer 150 meters (500 feet) thick. That's about the height of a 40-story building.

So how do the salts enter the ocean? Most of the salts actually come from the land. All natural water has some salts in it because rain dissolves salts from rocks and soil and washes them into streams. The stream water then flows into rivers that eventually empty into the ocean. Salts also enter the ocean from volcanoes and small vents on the ocean floor that release minerals, including salts, into the ocean water.

The entire global ocean is salty, but some parts are saltier than others. In the cold polar regions, some ocean water freezes into large chunks of ice. At Arctic and Antarctic temperatures, the ice is made of fresh water. Salt stays behind in the liquid water, making the surrounding water saltier than other areas of the ocean. In the hot tropics, ocean water **evaporates** into the warm air. As the water evaporates, it becomes a gas, and the salt stays behind in the liquid water. So this part of the ocean is saltier than the rest of the ocean.

Now think about what happens as fresh river water empties into the ocean. Even with carrying a little salt, the large amount of fresh river water dilutes the saltier ocean water near the mouth of the river. So that small part of the ocean is slightly less salty than other parts.

Ice floats in water, but not right on top! Like an ice cube, most of an iceberg floats below the surface.

Close to 50,000 icebergs can be seen each year in the Arctic Ocean.

A river in Norway dumps its silt, water, and dissolved salts into a sea that is part of the Arctic Ocean. The silt, or eroded land, forms the fan shape called a delta. The water and dissolved salts mix into the ocean.

Waves

"WATCH OUT!"

A giant wave curls over and crashes down upon a surfer. Then, seconds later, the surfer emerges from the barrel, the tunnel of water that forms as the top of the wave curls over. The skilled surfer rides the wave all the way to shore.

Surfers know all about waves. Most waves are caused by wind, or moving air. The blowing winds drag across the surface and pass energy to the water.

You might say the wind energizes the ocean as the energy passes through the water as waves. The stronger the winds

In 2012, a surfer rode a record-breaking 24-meter (78-foot) wave off the coast of Nazaré, Portugal.

are, the bigger the waves are. Imagine a storm at sea with fierce winds that kick up some serious waves. The energy might take many hours to reach the coast, but when it does, the waves become walls of water taller than houses.

Now imagine that you're out on your surfboard watching the sea. Wave after wave rolls beneath you causing you to bob up and down. The water around you bobs up and down too. The waves move forward, but the things on the water just move up and down.

Okay, time to catch a wave. You paddle forward and then stand up as the water swells beneath your board. It's a nice-sized wave, about half your height, but that's about to change.

The wave has grown as tall as you and now the wall of water towers over you. What's going on? The water is shallow enough that the bottom of the wave rubs against the ocean floor, which slows the wave but also pushes it higher.

You can feel the powerful wave breaking! The bottom of the wave slows down more than the top does, so the top leans forward, curls, and topples over. That's about the only time that a wave actually pushes water forward. It jolts you forward too, but you keep your balance. What a fantastic ride!

Sea stars, barnacles, and other tide pool organisms can live both in and out of water.

Nearly 4,000 kinds of fish dart and glide among the reefs of the ocean.

Animals in the open ocean swim for miles in search of food.

Ocean Life

From shallow shores to the deepest waters, Earth's ocean is home to organisms of every color, shape, and size. Some organisms are swimmers, such as whales, dolphins, octopuses, and fish. Bottom dwellers, such as lobsters, corals, and sea stars, live and walk, or crawl, on the ocean floor. Jellies and many kinds of microscopic organisms are floaters, staying at or near the surface. Seaweeds might float or be anchored to the bottom.

All ocean organisms are adapted to life in the ocean's salty water. For example, to avoid having too much salt in their bodies, fish pass extra salt through their skin and gills.

There are many ocean environments, each with its own unique mix of life. Tide pools are rocky depressions along shore. They are underwater during high tide and remain filled with water during low tide. The watery solution in tide pools becomes more salty as water evaporates and less salty when the tide comes in or when it rains.

The greatest variety of ocean life is found on or near coral reefs. Ant-sized animals called corals produce limestone coverings that build the stony reefs in tropical waters.

Some reef visitors, such as sharks and dolphins, are more at home in the open ocean. Compare the organisms that live in the environments shown here. This is just a tiny fraction of the huge variety of life found in the world's salty ocean.

A mouthful of seawater can contain hundreds of thousands of tiny animals and plantlike organisms. Millions of bacteria live there, too. Spit it out!

In the Deep

In 1977, a scientist and a pilot slowly descended into the Pacific Ocean in a small research submarine called *Alvin*. Another scientist kept track of *Alvin's* progress from the mother ship on the surface. The goal of their expedition was to find out if some of the ocean's salt was coming from volcanic vents on the seafloor. No one had ever explored these vents. This was their chance, and it turned out to be the discovery of a lifetime.

As *Alvin* approached the ocean floor, the pilot called out, "There's clams out here!" The clams were giant—each bigger than a football. There were also anemones, crabs, mussels, fish, and colossal worms that were taller than a doorway. The worms had red tips that extended from slender white tubes.

One of the scientists later described the worms to her colleagues over the phone. "I asked, 'Hey, can you biologists tell us what these things are?' And they said, 'What? We don't know what that is. Hold everything!'"

The scientists had discovered creatures in the most unlikely place. How could all this life be flourishing nearly 3 kilometers (2 miles) beneath the waves, in total darkness? Without sunlight, how did the organisms get energy?

The nearby volcanic vents spew out a mixture of salts, minerals, chemical gases, and extremely hot water. One of these gases is food for bacteria that live inside the tube worms and other animals. The bacteria use the gas to produce chemicals that the animals use as food.

Since 1977, scientists have discovered many vents and the life around them. *Alvin* has been joined by dozens of subs and other high-tech tools for exploring the ocean. Maybe one day, you'll take a deep-sea voyage and join in the thrill of discovery.

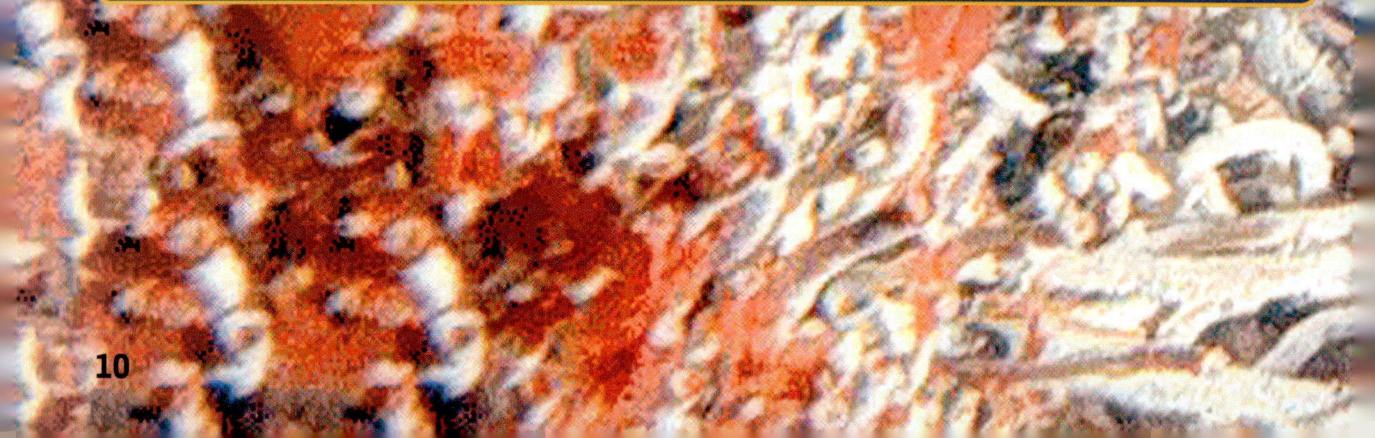

The black plumes, rich in chemicals such as sulfur, iron, zinc, and copper, measure over 350°C (662°F)!

The bright light shining on these strange animals comes from the submarine.

Check In Why are some parts of the ocean saltier than others?

GENRE **Process Article**

Read to find out how salt is harvested from the ocean.

Salt from the Ocean

by Judy Elgin Jensen

When you reach for a saltshaker in a restaurant, you might pick up a container labeled "sea salt." Sea salt is sold in grocery stores, too. What exactly is sea salt?

Sea salt is salt that comes from the sea, or ocean. The salt is harvested after seawater **evaporates** from shallow pools, leaving the salt behind. Table salt, the salt in most saltshakers, comes from underground mines.

Table salt and sea salt are not that different. They both are made of the particles sodium and chlorine, that combine to form sodium chloride. Many sea salts are left in large grains because coarse salt doesn't stick together like table salt can. You can use a grinder to break the sea salt into smaller grains. Sea salt also has tiny amounts of minerals that give the salt different flavors and colors. Chefs use these differences to enhance various foods. Here are just a few of the many kinds of sea salts available today.

Fleur de Sel That's French for "Flower of Salt." It's the best of the salt harvested in certain regions of France. Flavors vary slightly depending on the region.

Black Lava Sea Salt This black salt is produced on the Hawaiian island of Moloka'i. It is rich in minerals and has a slightly nutty flavor. It is colored with charcoal.

Chipotle Sea Salt Made with smoke-dried chipotle peppers and Pacific sea salt, this salt makes foods taste hot and spicy.

Red Alaea Sea Salt Pacific sea salt is combined with red clay from Hawai'i to color this salt. It is rich in minerals and is used to flavor and preserve food.

French Grey Sea Salt This coarse salt is harvested from the Atlantic waters off the coast of France. It smells like the ocean and is rich in minerals.

Apple Smoked Salt This flaky salt is smoked over a fire burning wood from apple trees. It absorbs the fruit flavor of the wood. It also contains minerals from the sea water.

Harvesting Sea Salt

Life in the ocean needs salt water to survive and people need some salt, too. A lot of the salt we use comes from the evaporation of salty seawater.

Sea salt is harvested in warm climates where the breezes are strong and steady. And of course, seawater must be nearby and plentiful.

The photos of these salt ponds show the basic process of harvesting sea salt on the island of Gozo, Malta, in the Mediterranean Sea. Let's see how it's done.

Step 1

First, seawater is pumped into shallow ponds, often made of local clay. Wind and sunlight slowly evaporate the water in the ponds. As more water evaporates, the remaining **solution**, called **brine**, gets saltier.

Step 2

The brine then moves through a series of ponds. After almost two years, the brine is so salty that solid salt crystals start forming out of the solution.

Step 3

Evaporation continues over time and more salt crystals form in the brine. When the salt layer is about 10 centimeters (4 inches) thick, the salt is broken up, scooped, poured into trucks, and taken to the wash plant.

Step 4

At the wash plant, the salt is churned in tubs to remove sand and dirt. The clean salt may be stored in mounds. The stored salt is then crushed, dried, and packaged. Then it's off to the store!

Importance of Salt

Salt is much more than a seasoning. Entire civilizations have been impacted by it. Hundreds of years ago, Venice, Italy, became an economic power by selling salt. In the late 1500s, the Dutch prevented war with Spain by blocking the transportation of salt from one of Spain's most important salt plants. Unable to sell their salt, Spain went bankrupt and could not fight a war.

A scale for measuring sea salt is a common tool at open-air markets. This market is in East Timor on an island north of Australia.

Sacks of local sea salts are on sale at a market on the Spanish island of Mallorca in the Mediterranean Sea.

Over the years, salt has been used as money. Soldiers of ancient Rome were paid partly in salt. In fact, the word *salary* comes from the Latin word for "salt."

How has salt had the power to build economies and influence wars? Salt is a substance your body needs, though in small amounts. It helps your muscles work, your blood flow, and your heart beat. Salt also slows the growth of bacteria, so it is used to preserve food. Salting fish and meats was important before refrigeration existed. Salting continues in many cultures today.

The popularity of sea salt has made this substance once again important to many local economies. A small sea salt works in Maine evaporates seawater in solar greenhouses to produce salt. A huge sea salt works outside San Francisco looks similar to those in Guatemala and around the world. Merchants compete for customers as the sea salt trend grows.

A crust of salt kept these fish from spoiling. They will be grilled and sold at a market in Bangkok, Thailand.

Check In Why is seawater pumped into shallow ponds rather than deep ponds?

GENRE Engineering Article

Read to find out about a solution to a freshwater supply problem.

Fresh Water from the Ocean

by Glen Phelan

EARLY DUBAI, 60 YEARS AGO

The dust is everywhere. It gets in your clothing, your hair, your eyes. Meanwhile, the sun is blazing and temperatures are soaring. That's just the way it is here in the Middle Eastern city of Dubai on the shores of the Persian Gulf. After all, this *is* a desert.

Children kick a ball in a sandy alley. When the ball comes your way, you see that it's not a ball at all, just a bunch of rags. This is a poor town where most people cannot read or write, and jobs are few. Many people live in tents and barely make a living fishing for sardines. But the children laugh and play as you tap the ball back to them.

You turn the corner and come nose-to-nose with a camel whose owner uses the animal to carry goods to one of Dubai's open-air markets. Colorful cloth draped over poles provides shade for people who stroll among the booths. The sweet aromas of cinnamon and cloves fill the air as do the sounds of people bickering over prices. You buy dried fruit and continue on your way.

A few minutes later you come to a small creek, which unlike most creeks, is salt water. It empties into the Persian Gulf, which is salt water too. Little rain falls, and there are no rivers and no big lakes to supply fresh water. So where do people get drinking water? A few deep wells provide water for the 30,000 people who live here. Still, fresh water is a scarce resource.

A saltwater creek divides the city. Everywhere else is dry, desert sand.

DUBAI, TODAY

Pop music blares while 1,200 stores invite customers inside. Now you are in the largest mall in the world, right here in Dubai. You see people wearing traditional Arabic clothing and teens wearing shorts and jeans. The Dubai Mall has an ice rink, an aquarium, a theme park, and movie theaters. Skyscrapers, including the world's tallest, line wide streets. People from all over the world stay in luxurious hotels. How did the dusty town become such a sparkling city?

Oil was discovered beneath Dubai's desert sands in the 1960s. It wasn't as much oil as in other parts of the Middle East, but it helped jump-start Dubai's economy. New wealth led to new businesses, industries, and schools. People came to build the modern city and tourists followed. The population grew to more than a million people.

As Dubai grew, it faced a major problem, a shortage of fresh water. People needed fresh water for drinking, washing, farming,

Peering up at the gleaming towers of glass and steel, or strolling through an indoor mall, it's hard to imagine the Dubai of 60 years ago.

and running factories and power plants. Scientists and engineers figured out how to get fresh water from salt water in a process called **desalination**.

Let's break apart the word *desalination*. The prefix *de-* means "to reverse or remove." The root word *saline* is Latin for "salt." The suffix *-ation* means "result of an action." So desalination is the action of removing salt.

There are several ways to desalinate salt water. Many desalination plants rely on evaporation and condensation.

When water **evaporates**, salt and other minerals stay behind, so the water that evaporates into water vapor is fresh. Many desalination plants catch the fresh water vapor and **condense** it, or turn it back into liquid. The condensed water is collected and piped to storage tanks. The Jebel Ali desalination plant in Dubai turns salty seawater from the Persian Gulf into drinkable fresh water.

Farmers in the Liwa Oasis irrigate with groundwater. The oasis is a four-hour drive from Dubai.

USING DESALINATED WATER

In parts of the Middle East, including Dubai, most water running out of the faucet came from the sea. Most of the drinking water for these countries comes from desalination plants.

Desalinated water is used in many ways. Industries use desalinated water to cool machinery and make products. Some farmers use desalinated water to irrigate fields, but most use **groundwater,** water that collects between rocks underground. Without desalination, cities along the coast would have to pipe groundwater over long distances. If people in the cities used more groundwater, less would be available for crops.

Desalination technology isn't perfect. It requires lots of energy and creates pollution. Most desalination plants

burn fossil fuels to heat the seawater, so to reduce pollution, engineers have designed desalination plants that use solar energy to heat the water.

Desalination also produces salty **brine** that can be unhealthy to nearby ocean life, so engineers designed a system in which the brine releases gradually. The brine is mixed more safely with the surrounding ocean water.

The water released back into the ocean can be warmer than the surrounding area, too. Warmer water is less dense than cooler water and holds less oxygen. This layer of warmer water can trap tiny ocean life and may not have enough oxygen for the organisms to survive. Scientists continue to study problems with desalination and design systems to provide solutions.

DESALINATION AROUND THE WORLD

Desalination is becoming more and more important worldwide. Saudi Arabia has 30 desalination plants that supply 70 percent of its drinking water. Other countries in the Middle East also rely on desalination for their drinking water. Desalination plants are operating around the world. In fact, more than 15,000 plants produce 65 billion liters (17 billion gallons) of fresh water every day! That's a lot! Yet that's only about two percent of all the fresh water people around the world use every day. Although desalination plants are

United Kingdom: London

The plant is powered with biofuel. It runs only in times of drought to supply 1 million people.

United States: Tampa

The plant supplies up to 10 percent of the region's needs. It helps conserve groundwater and reservoir stores.

Chile: Caldera

The plant ensures water needs for mining the rich mineral resources in the Atacama Desert. This desert is the driest location on Earth.

expensive to run, more of them are being built. Some are used only in times of drought while others sit far inland and extract salt from briny reservoirs and groundwater supplies.

Improved technology is making desalination more affordable and environmentally friendly. Maybe someday you'll work on the latest technology to help provide fresh water to a thirsty world.

India: Chennai

The plant supplies 4.5 million people. Water shortages result when erratic monsoons don't fill lakes and reservoirs.

Australia: Melbourne

The plant serves a growing population. Changing rainfall patterns mean reservoir stores can't keep up.

South Africa: Mossel Bay

The plant can supply all of the town's needs in case of a severe drought. Droughts are somewhat common here.

Check In What problems in the desalination process are engineers trying to solve?

GENRE Science Article

Read to find out why beaches are different colors.

THE OCEAN'S *Rainbow Beaches*

by Jennifer K. Cocson

Punalu'u Beach

Seawater is much more than just water. It's a **mixture** of water, salts, minerals, gases, and microscopic life. When that watery mixture breaks on shore, it washes over another mixture, the beach. Yes, a beach is a mixture. In every handful, you'll find sand, shells, and pebbles that you can easily pick out.

Where do these beach ingredients come from? The shells wash up from the ocean while the rocks and sand usually come from the land nearby. Rivers carry **eroded** bits of rock downstream and dump the rocky sediment near the coast. Waves smash the sediments against each other, breaking them into smaller pieces, and eventually into grains of sand. The crashing waves also erode, or wear down, the rocky shoreline, especially during storms. The waves constantly rework the sediment, forming a beach that is a mixture of the local rocky material.

Beach sand comes from nearby rocks, and rocks are all different colors. That means beaches come in a rainbow of colors, too! Let's take a look at some of the world's most colorful beaches.

Hawai'i is famous for its gorgeous beaches. Take Punalu'u Beach on the Big Island of Hawai'i, for instance. Its black grains are tiny bits of hardened lava. Hot lava flows down the volcano that forms the island. It cools upon hitting the ocean water and turns into black rock called basalt. The waves pound the basalt into bits of black sand that wash up to create this black sand beach.

The black sand of this beach in Hawai'i comes from the black volcanic rock that makes up much of the island.

What makes the sand at Pu'u Māhana Beach in Hawai'i green?

Pu'u Māhana Beach

On the west side of the Big Island you'll find Pu'u Māhana Beach, with sand grains that are like tiny green jewels. The olive-green color comes from the mineral olivine.

The cliffs around the beach are the remains of an old cinder cone, a hill formed by volcanic debris. These cliffs are loaded with olivine. Rain washes bits of olivine down the cliffs while waves erode the cliff and wash olivine on shore. Lighter particles are often swept out to sea and the heavier olivine grains are left behind.

The grains of olivine are too small to be valuable, but larger crystals of this mineral are the semi-precious gem called peridot.

Now hop over to Pink Sands Beach in the Bahamas. Can you guess where the pink sand comes from? It comes from the bright pink or red shells of microscopic animals that live among the nearby coral reefs. After the animals die, the waves smash the shells and wash the bits onto the shore. The shells mix in with the white sand to give the beach a pink color, especially in the wet sand at the water's edge.

Pink Sands Beach

The color of Pink Sands Beach in the Bahamas has a surprising source.

Rainbow Beach

Let's end our tour with Rainbow Beach, a colorful beach on the east coast of Australia. The name comes from the multi-colored sandy cliffs and dunes behind the beach. If you were to walk along this part of the coast, you might be able to count more than 70 shades of tan, yellow, orange, pink, red, blue, and brown.

A native legend tells of a spirit that took the form of a rainbow. After being killed in a fight, the spirit crashed into the cliffs, giving them their rainbow of color. The legend is an important tale in the native culture. However, scientists know that a mixture of different minerals is what gives the cliffs their colorful layers.

Erosion from wind and waves break apart the cliffs and forms piles of sand at the base. Wind and waves shift the sands and create an ever-changing colorful beach.

The sandy rainbows at Rainbow Beach in Queensland, Australia change often.

Look at all that color in the sandy cliffs of Rainbow Beach. How do you think these cliffs affect the appearance of the beach?

You can see colorful sands just by looking at Rainbow Beach. But as you dig into the sand, each scoop reveals a different combination of color. Talk about sand art!

The next time you are at a beach, observe what makes up this sandy mixture. Then try to figure out where the components of the mixture came from and why they look the way they do. Are there hills and cliffs nearby? What are the nearby rocks like? What kind of minerals do they contain?

Look at a map. Does a river empty into the ocean or lake close by? What kind of land did the river flow through on its long journey here?

Whew! Those are a lot of questions. Searching for the answers is where the adventure begins!

Check In How can you predict the color of a beach by looking at the nearby landforms?

Discuss

1. Tell about some of the ways you think the four pieces in *The World's Ocean* are connected.

2. Think about the process of getting salt from the ocean described in "Salt from the Ocean" and the process of getting fresh water described in "Fresh Water from the Ocean." Identify the cause and effect in each one.

3. Explain how the physical properties of the surrounding rocks and soils impact salt in "Salt from the Ocean" and sand in "Rainbow Beaches."

4. Cite evidence from "Our Salty Ocean" for why salinity varies in locations around the world.

5. What questions do you still have about the world's ocean? What would be some good ways to find out more information?